ROSEN Verified
NATURAL DISASTERS

TORNADOES

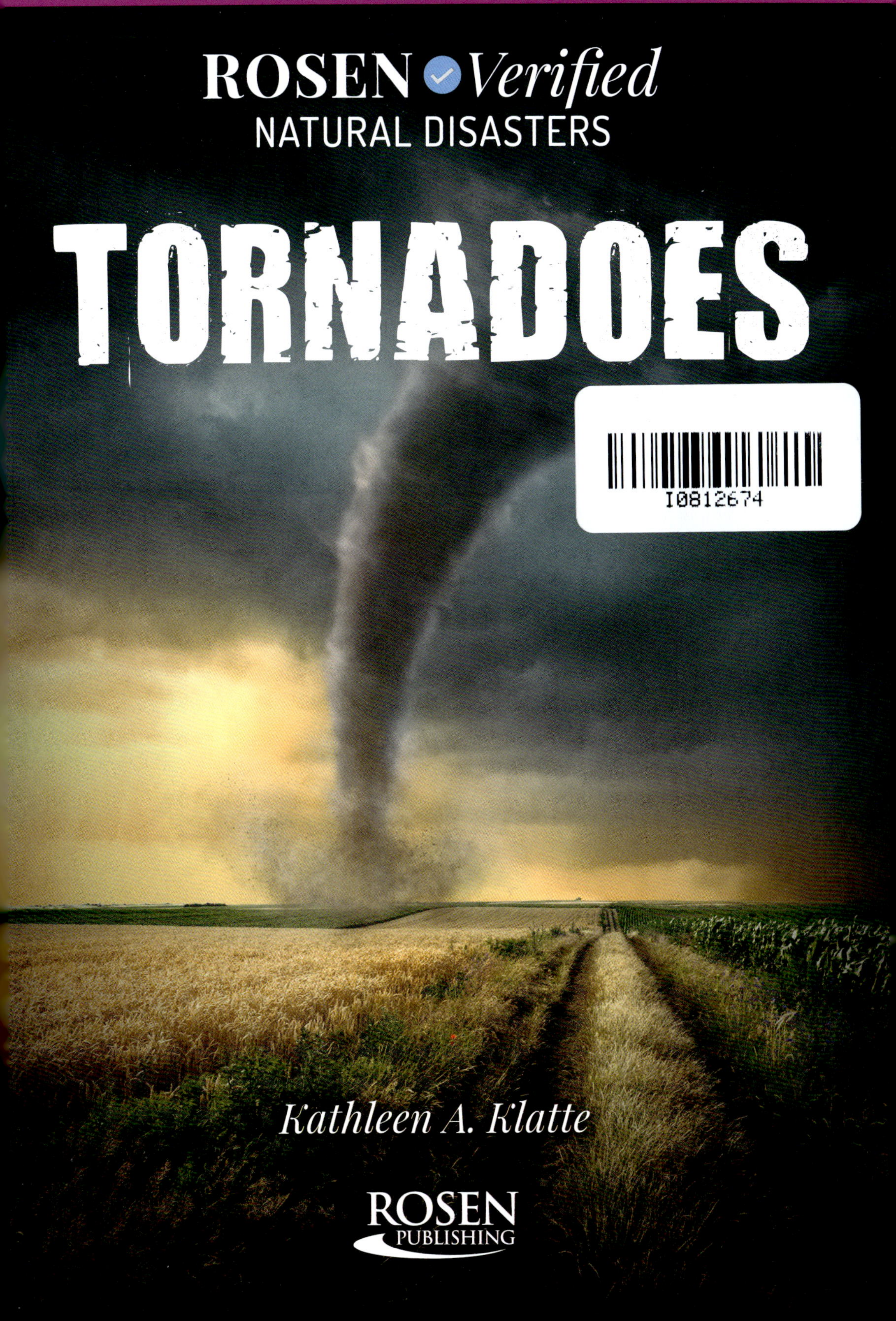

Kathleen A. Klatte

ROSEN
PUBLISHING

Published in 2023 by The Rosen Publishing Group, Inc.
2544 Clinton Street, Buffalo, NY 14224

First Edition

Editor: Greg Roza
Book Design: Michael Flynn

Photo Credits: Cover Rasica/Shutterstock.com; (series background) PinkPueblo/Shutterstock.com; pp. 5, 13 Cammie Czuchnicki/Shutterstock.com; p. 7 AP Photo/Pavel Rahman/Shutterstock.com; p. 9 Rainer Lesniewski/Shutterstock.com; p. 11 Matt Jeppson/Shutterstock.com; p. 12 Nick Uthe/Shutterstock.com; p. 15 NOAA OMAO; p. 17 John D Sirlin/Shutterstock.com; p. 18 sdecoret/Shutterstock.com; p. 19 Dave Weaver/Shutterstock.com; p. 21 (siren) Darryl Brooks/Shutterstock.com; p. 21 (alert) Matteo Benegiamo/Shutterstock.com; p. 23 Photoguru73/Shutterstock.com; p. 24 Sari ONeal/Shutterstock.com; p. 25 Noel V. Baebler/Shutterstock.com; p. 27 steve Hayeslip/Shutterstock.com; p. 28 Andrey VP/Shutterstock.com; p. 29 Andrey Armyagov/Shutterstock.com; p. 31 LeStudio/Shutterstock.com; p. 32 solarseven/Shutterstock.com; p. 33 Fouad A. Saad/Shutterstock.com; p. 35 Rob Atherton/Shutterstock.com; p. 37 Alan Budman/Shutterstock.com; p. 39 Alexey Stiop/Shutterstock.com; p. 41 https://commons.wikimedia.org/wiki/File:DOROTHY_und_D.O.T._3.jpg; p. 43 kirill_makarov/Shutterstock.com; p. 45 solarseven/Shutterstock.com.

Cataloging-in-Publication Data

Names: Klatte, Kathleen A.
Title: Tornadoes, Kathleen A. Klatte.
Description: New York: Rosen Publishing, [2023] | Series: Rosen Verified : Natural Disasters | Includes index.
Identifiers: LCCN 2022028995 | ISBN: 9781499469615 (library binding) | 9781499469608 (paperback) | 9781499469622 (ebook)
Subjects: LCSH: Tornadoes - Juvenile literature
Classifications: LCC QC955.2 .K53 2023 | DDC 551.553

Manufactured in the United States of America

CPSIA Compliance Information: Batch #CWRYA23. For further information contact Rosen Publishing at 1-800-237-9932.

CONTENTS

THUNDER AND LIGHTNING!

Few people like a thunderstorm. At best, it can ruin your plans for the day. At worst, there might be flooding or other damage, or harm. Sometimes, a thunderstorm can be the first sign of an even worse storm. A tornado is a type of **severe** storm that can form during a thunderstorm. Tornadoes can produce the strongest winds on earth.

Tornadoes are best known for their **distinctive** funnel-shaped clouds. Everyone's seen that in a movie. Although a real tornado won't whisk you off to Oz, it can do **tremendous** damage in a short time.

Tornadoes occur all over the world. Fortunately, many are too weak to cause much damage. Sometimes tornadoes happen in places where no one lives. When a tornado does touch down in a place where lots of people live, the results can be terrible.

Not every thunderstorm can turn into a tornado. A supercell thunderstorm, shown here, can cause a tornado.

Most tornadoes occur in the United States. Although there are warning systems and shelters, people die each year in tornadoes. Homes and businesses are damaged or destroyed.

Scientists study tornadoes to try and figure out how they happen. They want to be able to create better warning systems and shelters.

ALL OVER THE WORLD

Tornadoes can occur anyplace the right conditions exist. This includes most places around the world. South America, Europe, and Australia have all reported tornadoes.

The United States has the greatest number of tornadoes in the world. Most occur in the Plains states. This is because cold, dry air from Canada collides with warm moist air from the Gulf of Mexico there. These conditions are ideal for forming supercell thunderstorms.

The only part of the world with no reported tornadoes is Antarctica. The air is too cold and dry.

LATITUDE AND LONGITUDE

Latitude is the distance north or south of the **equator**. Longitude is the distance east or west of the **prime meridian**. Tornadoes generally occur in the mid-latitudes—not at the equator or poles.

Tornadoes in Bangladesh aren't common, but they can be deadly. This is because many people live there. Many houses aren't well constructed. A tornado in April 1989 killed 1,300 people in Bangladesh.

TORNADO ALLEY

The part of the United States where most tornadoes occur is called Tornado Alley. It extends from Texas through Oklahoma, Kansas, and Nebraska. The area averages more than 250 tornadoes per year. It's most active from April to June.

Texas has the most tornadoes each year, followed by Kansas. Seven of the most expensive storms since 1950 have occurred in Tornado Alley. Tornadoes also caused around 1,100 deaths in the area between the years 1950 and 2012.

PLAN AHEAD

People who live where tornadoes are common should have a plan to locate family members after a strike. It's best to contact a friend or family member in another area. It's also better to text and only use voice phone lines for **emergency** calls.

THE SOUTHEASTERN TORNADO ALLEY

There is another area of the United States that has tornadoes. It includes parts of Arkansas, Louisiana, Mississippi, Alabama, and Georgia. Besides the regular tornado season, this area can experience tornadoes in late summer or fall.

WHERE THE WINDS BLOW!

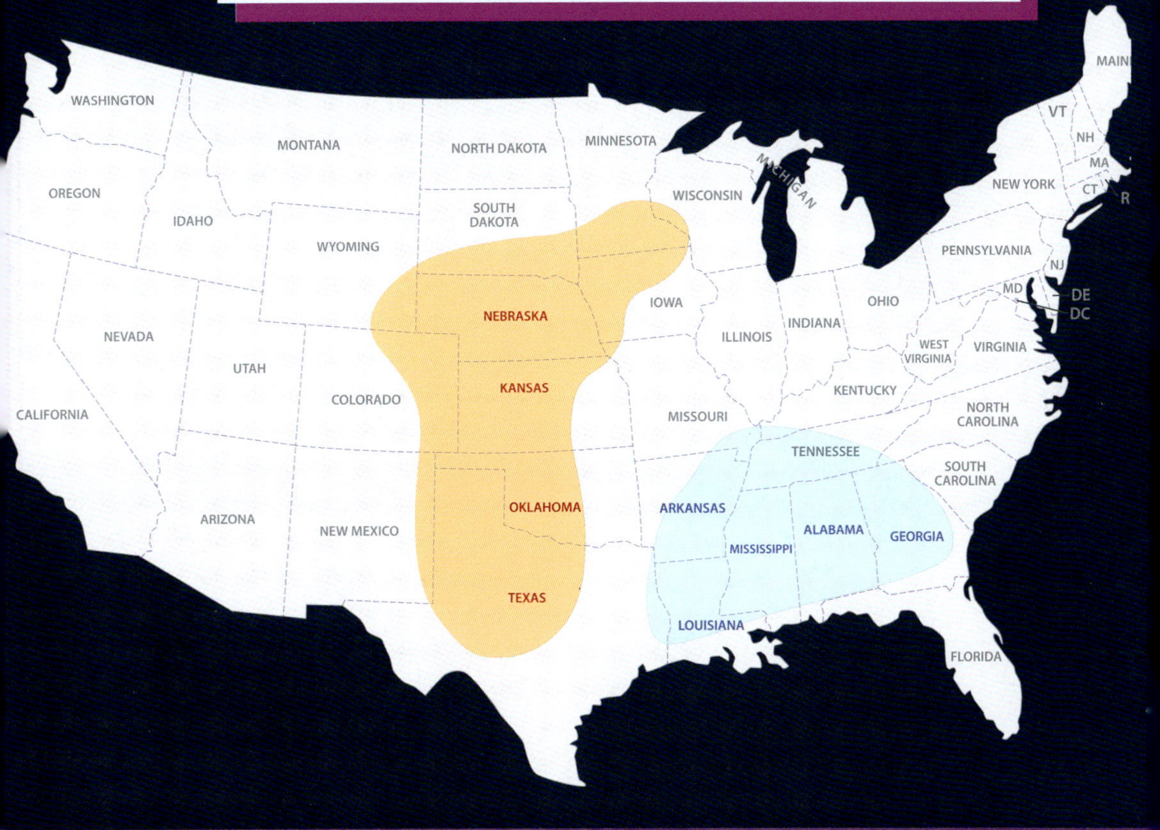

Most tornadoes in the Unites States happen in Tornado Alley because the conditions are perfect for forming severe thunderstorms.

TORNADO SEASON

In the United States, most tornadoes occur in spring and summer. About three-fourths of all tornadoes happen between March and July. May has the highest occurrence. However, there have been tornadoes reported all year long. Southeastern states have reported tornadoes in autumn.

People who live where tornadoes are likely to form must be **vigilant** all year long. They keep their emergency supplies up to date and pay close attention to weather reports.

DECLARING A DISASTER

A disaster is a sudden event that causes great damage or death. A state governor can apply for federal aid when a disaster is too big for local authorities to handle. Damage caused by a storm such as a tornado may be considered a major disaster. Once the president declares a major disaster, federal agencies can bring in supplies, equipment, and rescue workers.

VERIFIED

The Federal Emergency Management Agency (FEMA) is part of the U.S. Department of Homeland Security. FEMA helps Americans prepare for disasters. It also responds with aid and equipment, or gear, after a disaster. You can learn more at **www.fema.gov**.

The costliest tornado on record struck Joplin, Missouri, in 2011. It caused $2.8 billion in damage.

NOT JUST THE WEATHER

The science of weather and the **atmosphere** is called meteorology. Scientists who study weather are called meteorologists. You might be familiar with a TV meteorologist giving a weather report, but they do much more.

Some meteorologists are storm chasers. They drive vehicles loaded with portable equipment. They may drive as close to a storm as they can to get data. This is very dangerous.

I'M ON TV!

A TV weather presenter does most of their work in front of a camera. Their job is to communicate information to the public. They may or may not hold degrees in meteorology. A TV weatherman might expect to earn about $56,000 to $68,000 per year.

FAST FACT

Many universities offer meteorology majors.

Some meteorologists work in labs. They use **radar** and **satellites** to collect as much data, or facts, as possible. They use computers to make simulations, or models of storms. They study data about storms to find out how they work.

NOAA

The National Oceanic and Atmospheric Administration (NOAA) is part of the U.S. Department of Commerce. Its job is to monitor the weather and atmosphere.

The National Weather Service is part of NOAA. It issues weather **forecasts** and storm watches and warnings.

NOAA uses cutting-edge **technology** to study weather. It learns as much as possible about how storms form and work. Its forecasts help people know when it's safe to travel. Companies depend on them to know if it's safe to transport goods or hold events.

AT THE HEART OF IT ALL

NOAA has offices in Norman, Oklahoma. This is in the heart of Tornado Alley. Scientists work there to study tornadoes. They want to know as much as possible about how they work. With more information, they can design, or plan, better warning systems.

VERIFIED

You can check out NOAA's official website at **www.noaa.gov**.

NOAA has research vessels to study the oceans. It has planes that can fly right into **hurricanes**. It also owns and operates satellites.

HOW DO TORNADOES FORM?

Most tornadoes form during severe storms called supercell thunderstorms. These storms have strong winds and lots of moving air. A tornado forms when a mass of cold, dry air lies over a mass of warm, moist air. The warm air rises and cool air falls. It can start to spin. To be called a tornado, the column of spinning air must reach from the storm clouds to the ground.

Not all supercell thunderstorms spawn, or produce, tornadoes. Scientists aren't sure why.

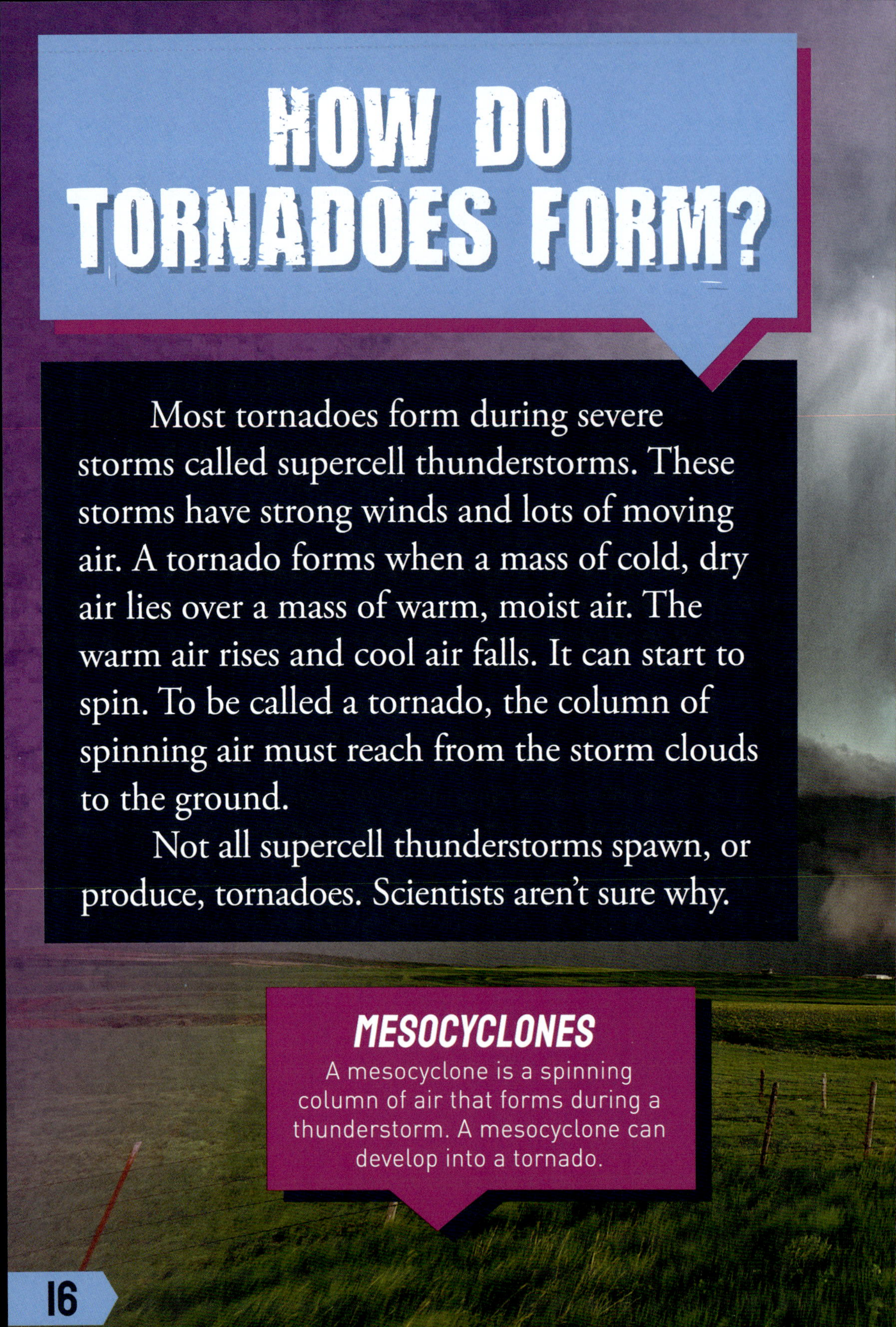

MESOCYCLONES

A mesocyclone is a spinning column of air that forms during a thunderstorm. A mesocyclone can develop into a tornado.

FAST FACT

Tornadoes aren't named. They're often called by the name of the place they struck or caused the most damage. Hurricanes are the only major storms given official names.

Supercell thunderstorms are very dangerous, even if they don't spawn tornadoes. They feature strong winds and lots of lightning. Sometimes there's damaging hail.

WHY DO THEY CAUSE SO MUCH DAMAGE?

Tornado damage is caused by high winds. The extreme winds of a tornado can uproot trees and lift vehicles from the ground. Tornado winds can also remove roofing and siding from buildings. Sometimes these large objects can impact things like buildings. This can cause even more **destruction**.

After a tornado, people want to see what they can **salvage** from their homes. It's important to wait for the authorities to declare the area safe. There could be gas or sewage leaks. The buildings might also be too unstable to search safely.

Winds can also pick up all sorts of small debris, or the remains of something that has been destroyed. Things like nails and bricks can cause plenty of damage when they become **airborne**.

The storm that spawns a tornado is also dangerous. Supercells can produce dangerous lightning, hail, and **flash floods**.

DEADLY WINDS

The deadliest tornado on record is the Tristate Tornado of 1925. It killed 695 people in Missouri, Illinois, and Indiana. The deadliest tornado since 1950 hit Joplin, Missouri, in 2011. It killed 158 people.

WARNING SYSTEMS

Tornadoes can happen very quickly. People who live where they are common need to be alert. A tornado watch means that the right kind of weather exists for a tornado to form. A tornado warning means that a tornado has been seen.

Once a warning is issued, it's very important to go to the nearest safe shelter. Listen carefully to directions on the TV or radio. If you're in a public place, directions will probably be given over the speaker system.

TORNADO ALERT APPS

The Red Cross, NOAA, and several weather websites offer tornado alert apps. If you live where tornadoes are common, getting alerts sent right to your phone can help keep you safe.

✓ VERIFIED

The National Weather Service issues official watches and warnings. It's the most trusted source for weather information. Its website is **www.weather.gov**.

If you can hear sirens or announcements, it's important to pay attention. Don't take time to call 911. Get to shelter right away.

TORNADO DRILLS

If you live where tornadoes are common, it's important to know what to do if one is sighted. Tornadoes form and move very quickly. There's not much time between a warning and a strike.

Once you hear a siren or other alert, seek shelter immediately. Schools hold tornado drills. Families should too. It's important to know where to go and how to get there quickly. Families should also have a plan to check in if they're not together when a storm strikes.

FAMILY TORNADO DRILLS

Your family should practice what to do in an emergency. It's very hard to think when sirens are blaring. The lights may be flickering or you might be asleep when the storm hits. Having a good plan that you've practiced can help keep your family safe.

In a school tornado drill, students use the stairs to go to the lowest floor of the building. They kneel facing a strong interior, or inside, wall and cover their heads.

EMERGENCY KITS

Your family should have an emergency kit prepared for extreme weather. It should include enough supplies for three days. Here are some things you might include:

- Water
- Food that doesn't need cooking or refrigeration
- First aid kit
- Pet supplies
- Medicine
- Baby supplies

STORM SHELTERS

If you live where tornadoes are common, you'll see signs for public storm shelters. Homes often have storm cellars. But what if you live someplace where tornadoes aren't common?

Tornadoes can happen anywhere, at any time of year. You need to take shelter even if there's no special place to hide. Take the stairs to the lowest floor of the building. Look for a small interior room, like a bathroom or closet. A short interior hallway can also work.

Places where tornadoes are common will have well-marked shelters. If you can't get to a shelter, go to a small inside room, like a closet or bathroom.

Homes where tornadoes are common often have storm cellars. These are underground bunkers. People go inside to be protected from extreme winds and flying debris.

THINGS TO AVOID WHEN SEEKING SHELTER

WHAT	WHY
Windows and glass doors	Glass can shatter
Long straight corridors	Can become wind tunnels
Gyms or auditoriums with large, flat roofs	Roofs can collapse, or fall in
Rooms with outside walls	Flying glass, building collapse
Mobile homes	Can be picked up by extreme winds

HOW BIG?

It's hard to say how big a tornado is. They form and disappear very quickly. Scientists tend to judge their size by the path of damage left by a storm. The width of the path can vary from a few yards to a mile (1.6 km). The length of the path can be just a few yards or 100 miles (161 km). The same tornado can also change size and **velocity** rapidly.

Some people think "fat" tornadoes are stronger. That's not necessarily true. Small tornadoes can be very strong. They can cause widespread damage in a small area. A large tornado might be weaker, but spread damage over a large area.

RECOVERING FROM A TORNADO

Recovering from a tornado can mean lots of things. People may have lost their homes and possessions. Towns might have to figure out how and where kids will go to school. If stores were damaged or destroyed, people may have to drive to get supplies.

A tornado that hit El Reno, Oklahoma, on May 31, 2013, was 2.6 miles (4.2 km) wide. It's considered the widest tornado on record.

HOW FAST?

It's hard to measure wind speeds inside a tornado. Strong tornadoes can destroy measuring equipment. Readings can be taken inside weaker tornadoes. Radar is used to measure wind speeds above ground level.

The fastest wind speed recorded in a tornado was 318 miles (511 km) per hour. This happened in Oklahoma City on May 3, 1999.

A typical tornado can move 10 to 20 miles (16 to 32 km) per hour. Some have moved as fast as 60 miles (96.5 km) per hour.

Scientists don't know the answers to lots of questions about tornadoes. It's hard to get readings about a storm that picks up equipment and wrecks it.

NASA

The National Aeronautics and Space Administration—NASA—studies tornadoes and other extreme storms with satellites. Satellites are above the storm, so they don't get damaged. They can track the damage from a storm. This is helpful for first responders on the ground.

VERIFIED

NASA does far more than study tornadoes. They are a part of the U.S. government that studies and works in space! Visit the NASA website to learn much more about the organization, what it does, and the many amazing discoveries they have made about Earth and the universe: **www.nasa.gov**.

THE FUJITA SCALE

The Fujita scale was created in the early 1970s. It ranked tornadoes on a scale of 0 to 5. It was based on the damage they did. Wind speeds were estimated, or guessed at, based on the damage too.

In 2007 it was replaced by the Enhanced Fujita scale (EF scale). The EF scale considers many different factors. It takes into account the kinds of objects exposed to the wind. For example, winds of the same strength might destroy a mobile home, but damage a brick building.

THEODORE FUJITA (1920–1998)

The original Fujita scale was developed by Dr. Tetsuya Theodore Fujita of the University of Chicago. He was best known for his study of thunderstorms, tornadoes, and hurricanes. He discovered macrobursts and microbursts. These are wind events that are dangerous for pilots.

FUJITA SCALE

RATING	WIND SPEED	DAMAGE
EF0	65–85 mph	minor roof, branches
EF1	86–110 mph	broken windows
EF2	111–135 mph	roofs off, large trees
EF3	136–165 mph	homes damaged
EF4	166–200 mph	homes leveled
EF5	200+ mph	incredible damage

A "significant" tornado is one that's rated EF2 or higher. This is a storm that can remove roofs from houses or destroy mobile homes. It can uproot trees or tip over railroad cars.

WHY DOES IT LOOK LIKE THAT?

Tornadoes are known for their distinctive shape. This shape is called a funnel cloud or condensation funnel. The funnel cloud is the column of water droplets that goes from the storm cloud to the ground. It's visible because the rotating wind picks up dirt and debris.

The funnel cloud is the small center of the storm that's visible to human eyes. But the tornado can extend much farther than the funnel.

The funnel cloud is the most distinctive feature of a tornado. It's only about 1/10 of the whole storm.

WHAT DIRECTION DO TORNADOES SPIN?

Tornadoes in the Northern **Hemisphere** often spin counterclockwise. In the Southern Hemisphere, they often spin clockwise. There are cases of tornadoes spinning in either direction.

THE CORIOLIS FORCE

Coriolis force is an effect of Earth's rotation. Objects in the Northern Hemisphere are turned to the right. In the Southern Hemisphere, they're turned to the left. Large storm systems like hurricanes are strongly affected by the Coriolis force. Tornadoes are too small for it to have much effect.

WATERSPOUTS

Waterspouts are tornadoes that form over water. They're common off the Florida coast. They can form over ocean water or fresh water. Waterspouts are considered weaker than tornadoes, but they can still sink ships.

In July 2006, a waterspout formed on the Hudson River, near Tarrytown, New York. It came ashore as an F2 tornado. It caused power outages and destroyed a store. A lot of people didn't take warnings seriously. Others didn't know what to do.

DEADLY WATERSPOUT

The deadliest marine tornado event in the United States occurred in 1814. The ship *Alligator* sank near Charleston, South Carolina. Twenty-five members of a forty-man crew were killed.

FAST FACT

The National Weather Service issues special marine warnings if waterspouts are likely.

Official tornado data has been recorded since 1950. Waterspouts aren't recorded as tornadoes unless they come ashore.

A WORLD OF WIND!

There are lots of different kinds of weather events with high winds. The words "cyclone" and "tornado" were once used for the same kinds of storms. Now meteorologists use the term "cyclone" to mean large storm systems like the ones that become hurricanes.

A gustnado is a small **whirlwind**. It's much weaker than tornadoes but can still do damage.

A derecho is a group of thunderstorms with strong straight-line winds. It can be very powerful. People might be told to seek shelter as they would for a tornado.

MICROBURSTS

A microburst is a violent **downdraft**. It causes winds to shift suddenly. Microbursts are a hazard to planes that are taking off or landing.

WHY WAS OUR FLIGHT CANCELED?

Commercial airline flights may be delayed or canceled for winds of about 45 miles (72 km) per hour. That may not sound very fast, but it's all about the direction the wind is coming from. Planes generally take off with the wind behind them. They land into the wind. Wind coming from the side is dangerous.

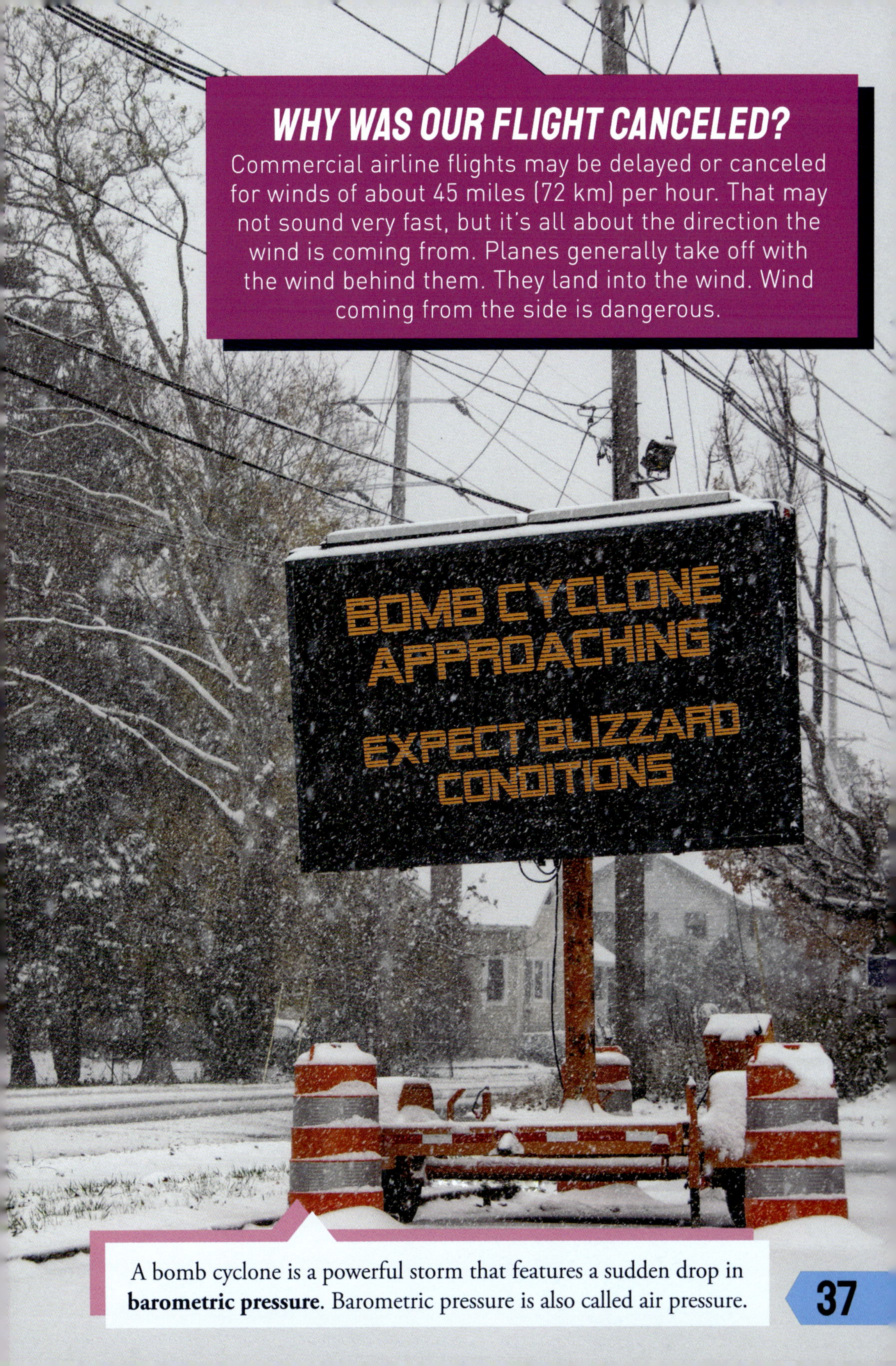

A bomb cyclone is a powerful storm that features a sudden drop in **barometric pressure**. Barometric pressure is also called air pressure.

THE 2011 SUPER OUTBREAK

The 2011 Super Outbreak is the largest outbreak of tornadoes on record. It occurred April 26 to 28, 2011. More than 300 tornadoes were recorded across 15 states. They caused about 340 deaths.

Alabama was the hardest hit. The state recorded 230 deaths and more than 2,000 injured people. Tuscaloosa was hit by an EF4 tornado.

An outbreak just days earlier, from April 14 to 16, spawned 155 tornadoes and claimed about 40 lives. Altogether, 817 tornadoes were recorded in April 2011. This is the most in a month since official records began in 1950.

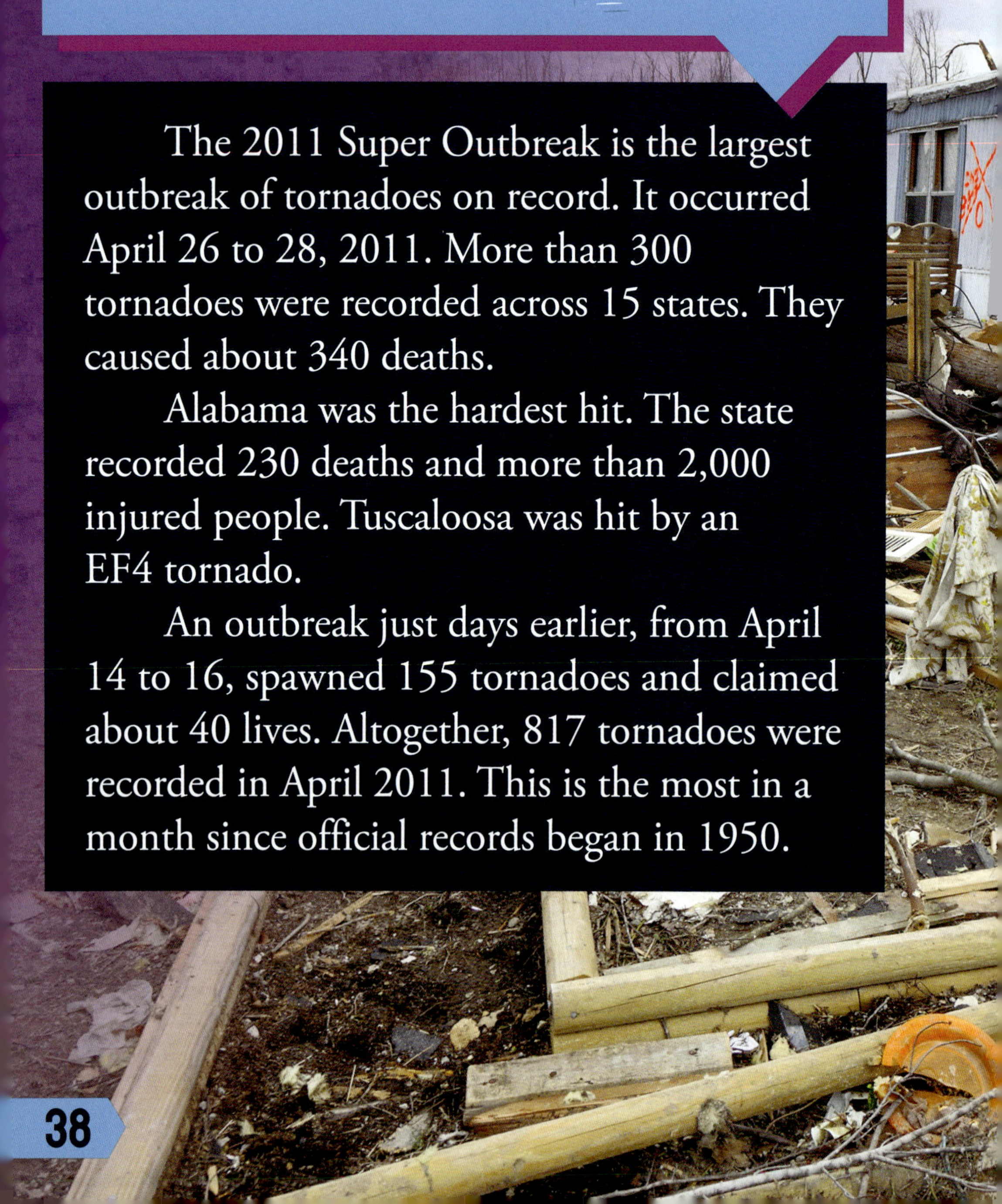

VERIFIED

The American Red Cross provides emergency aid for the public. It's often one of the first groups to respond to natural disasters. You can learn more about it at **www.redcross.org**.

Some people think that tornadoes actually "target" trailer parks, but that's not true. Mobile homes, or manufactured homes, are built at a factory and moved to a location. They are not as strong as other types of homes. Because of this, they tend to be damaged more in tornadoes than some other kinds of homes.

ABOUT THE MOVIES...

Twister (1996) is a movie about scientists who study tornadoes. It's not entirely true-to-life. Some details were made up for the movie to make the story more exciting. However, the actors spent time at the NOAA National Severe Storms Laboratory (NSSL) with real meteorologists.

The Day After Tomorrow (2004) is a movie about the effects of global warming. In the film, the melting of polar ice caps releases too much fresh water into the ocean. This changes ocean currents that influence climate. All sorts of extreme weather follow. Los Angeles, California, is leveled by tornadoes. While climate change is very real, it simply doesn't happen as quickly as the movie shows.

SORRY, HOLLYWOOD...

Real tornadoes are capable of moving houses from their foundations, but not transporting them to a magical land like in *The Wizard of Oz* (1939). They've also been known to pick up small fish and marine animals and drop them on land...but not great white sharks.

The Dorothy and D.O.T. devices in *Twister* were based on a real-life NOAA project. The TOtable Tornado Observatory (TOTO) was a barrel of tools designed to be placed in the path of a tornado. It was retired in 1987.

TORNADO MYTHS

There are a lot of myths about tornadoes. Most have been disproven by science. Some are dangerous to believe.

If a tornado has been sighted nearby, don't take time to open windows "to equalize pressure." A change in air pressure caused by tornadoes won't shatter glass. The impact of flying debris does. It's more important to use your time to get to shelter.

The southwest corner of a building isn't the safest, as some people believe. The best places are small, interior bathrooms, closets, or similar rooms.

THE BEST WAY TO KEEP SAFE

Green clouds aren't an accurate indication of an approaching tornado. A funnel cloud isn't always visible. Sometimes you can't hear a tornado until it's too late to get to shelter. The best way to stay safe is to listen carefully to official weather warnings.

A bridge or overpass isn't a good place to seek shelter. It can become a wind tunnel. Cars and large debris can be blown underneath by the tornado. You also run the risk of being caught in a flash flood.

WINDS KEEP BLOWING

Tornadoes are violent, dangerous storms. They can form quickly and cause terrible destruction in a short amount of time. But the more we know about them, the safer we will be when they occur.

Meteorologists will continue to study tornadoes. They'll continue to learn more about how they form and function. They'll use the data they collect to design better warning systems. They use portable equipment and drive as close as they can to tornadoes that are forming. They use new technology to gather more data. Unfortunately, it's hard to study a storm that destroys scientific instruments!

Right now, a tornado watch can be declared hours in advance, but a warning might only give people minutes to seek

Tornadoes are one of Mother Nature's most awe-inspiring sights. They can be beautiful from a safe distance.

shelter. Scientists want to be able to extend that time and save lives. For now, the best way to live safely with tornadoes is to pay close attention to official announcements. Be alert to warnings and seek shelter at once if a warning is declared.

Does chasing after a storm that can level brick buildings sound exciting? Maybe a career as a meteorologist is in your future!

GLOSSARY

airborne: Moving or being carried through the air.

atmosphere: The mixture of gases that surround a planet.

barometric pressure: The weight of air above an area of Earth's surface.

destruction: The act or process of damaging something so badly that it no longer exists or cannot be repaired.

distinctive: Having a special look that is easy to identify.

downdraft: A current of gas traveling down (such as air during a thunderstorm).

emergency: An unexpected situation that needs quick action.

equator: An imaginary circle around the middle of Earth that is the same distance from the North Pole and the South Pole.

flash flood: A local flood of short duration generally resulting from heavy rainfall.

forecast: To say that something will happen in the future after looking at the information that is available.

hemisphere: A half of Earth.

hurricane: An extremely large, powerful, and destructive storm with very strong winds that occurs especially in the western part of the Atlantic Ocean.

prime meridian: An imaginary line that runs from the North Pole to the South Pole through Greenwich, England.

radar: A device that sends out radio waves for finding out the position and speed of a moving object (such as an airplane).

salvage: To save or rescue possessions in danger of being lost.

satellite: A machine that is sent into space and moves around Earth, the moon, the sun, or a planet.

severe: Very bad, serious, or unpleasant.

technology: A method that uses science to solve problems and the tools used to solve those problems.

tremendous: Very large or great.

velocity: The speed of an object in a certain direction.

vigilant: Watchful for danger.

whirlwind: A very strong wind that moves in a spinning or swirling motion that can cause damage.

INDEX